THE SILENT ELEMENT

THE SILENT ELEMENT

On the Intelligence Hidden in Water

Approved by the National Media Council of the United Arab Emirates
No.: MC-01-01-6211747
Date: 13 March 2026

ISBN 978-9948-63407-2 (paperback)

Cover design by Sıla Eser
Interior design by Asia Bizior

First edition
2026

Published by Aeon Origins Press
Dubai, United Arab Emirates

CONTENTS

"Nothing is softer or more flexible than water, yet nothing can resist it."

— LAO TZU

PROLOGUE

Before We Learned to Look

In 1968, astronaut William Anders took a photograph from lunar orbit that quietly reshaped how humanity saw itself. A blue sphere in darkness. No borders. No cities. Just rock, water, and cloud. They called it *Earthrise*. For the first time, people noticed what had always been true: the dominant color of their planet — the reason it was visible at all — was water.

We live on a water planet. And perhaps that is exactly why we stopped seeing it.

There is a number I encountered several years ago that I have not been able to forget.

2.2 billion.

That is the number of people on Earth who, according to the United Nations, lack reliable access to safe drinking water. Not in some historical past. Now. Today. While you hold this book.

I had known, in the vague way most of us know, that water scarcity existed. I had read the headlines. But there is a difference between knowing a fact and allowing it to reach you — the way a sound you have been hearing for years suddenly resolves into a voice.

2.2 billion people. Roughly one in every four human beings alive.

Once that number became real to me, something else became impossible to ignore: the world I live in — the world of taps and bottles and water delivered without thought — is not normal. It is an exception. A narrow, historically recent exception that most of the planet does not share.

And then I encountered a second fact, smaller in scale but sharper in its strangeness:

In dozens of countries across the world, one liter of drinking water costs more than one liter of gasoline.

Read that again.

The substance without which a human body fails in three days — the biological fuel of every cell, every organ, every thought — is, in much of the world, more expensive and less accessible than the fuel we pump into machines.

We have built a civilization that values the energy of engines above the energy of bodies. Not by ideology. By structure. By the quiet accumulation of economic choices that no one paused long enough to question.

This is not a book born from that outrage alone, though the outrage is real. It is born from what happened next — from the questions that followed once I began to look at water not as background, but as foreground.

Because once you start paying attention, other dissonances appear.

We live in an era of extraordinary health sophistication. We track sleep in ninety-minute cycles. We calibrate macronutrients to the gram. We measure heart rate variability, optimize recovery protocols, sequence genomes to personalize medicine. The modern pursuit of health has become, in many ways, a pursuit of precision.

And yet the substance that constitutes sixty percent of the human body — the medium in which every biochemical reaction occurs, the carrier that delivers minerals to cells, the environment in which the brain operates — remains, for most people, an afterthought.

Drink more water. Stay hydrated.

That is the extent of the conversation.

But here is what that conversation leaves out: water is not generic. It never was.

Every water carries a geological signature — a mineral composition shaped by the specific rock, pressure, and time through which it traveled. That signature is not decorative. The human body reads it. Our kidneys respond to it. Our cells process it differently depending on what it contains. For three hundred thousand years, Homo sapiens drank water whose character was defined by geography — diverse, variable, shaped by the landscape through which it moved.

Then, in roughly a century, we standardized the water. We filtered it. Purified it. Made it uniform. We reduced an infinite spectrum of geological identities to a single anonymous commodity. We did this for excellent reasons — to eliminate disease, to scale infrastructure, to protect public health. And we succeeded. Access to safe water is one of the greatest engineering achievements in human history.

But in solving one problem, we silently obscured another.

Not the water itself — but our relationship to it. In the pursuit of safety and scale, we lost the ability to perceive water as something with character, with composition, with a biological identity that matters. We stopped asking what our water contained, where it came from, what it was doing inside us. We replaced a rich, varied, geologically specific relationship with a single instruction: drink more.

This book is about what was lost in that simplification.

It is about the intelligence hidden in water — not mystical intelligence, but geological intelligence: the compositional identity that water acquires over years and decades underground, and that the human body has spent its entire evolutionary history learning to interpret.

It is about the recognition that different waters serve different biological purposes — that a dense, calcium-rich spring water and a light, naturally soft glacial water are not inferior and superior versions of the same substance, but distinct biological inputs, each suited to different contexts, different bodies, different moments.

And it is about a shift that I believe is coming, whether or not this book accelerates it: the recognition that hydration is not a question of volume, but of context. That the central question is not "Am I drinking enough?" but "Does the water I drink match what my body needs right now?"

This is not a wellness argument. It is a civilizational one.

Because a world that has learned to value the fuel of its machines above the fuel of its bodies has lost track of something fundamental. And a species that has spent three hundred millennia in conversation with geologically diverse water — and then, in a single century, reduced that conversation to a number — should probably ask what was simplified away.

These pages are an attempt to answer that question.

Not with alarm. Not with prescriptions.

With the kind of careful attention that the subject has always deserved and rarely received.

After this book, you may not change anything about how you drink. That is your right and it is not my intention to dictate.

But I suspect one thing will shift.

The next time you pick up a glass of water, you will pause — even for a fraction of a second — and wonder what it contains. What it traveled through. What it is carrying into you.

That pause is where the conversation begins.

And once it begins, it does not easily stop.

CHAPTER 1

The Hidden Intelligence of Water

In the small town of Évian-les-Bains, on the southern shore of Lake Geneva, there is a spring from which people have been drinking for over two centuries. In 1789, the Marquis de Lessert, walking through the foothills of the Alps, stopped at a natural spring. He drank from it for several weeks — and noticed that the kidney ailment that had troubled him for years began to recede.

The Marquis did not know the word "mineralization." He could not explain what this water contained. But his body recognized the difference.

Two hundred years later, we know that the water from this spring travels through Alpine glacial rock for roughly fifteen years, dissolving calcium, magnesium, and bicarbonates along the way. We can measure every microgram. And yet most of us still treat water as though it were without character.

You drink it every day.

Without thinking. Without questioning. Almost never wondering what it actually is.

Water is the most familiar substance in your life. And, quite possibly, the most underestimated.

We have learned to track sleep, count steps, measure heart rate variability, calibrate nutrients, study biomarkers, and invest in longevity. The modern person seeks to manage their body with a precision that would have seemed impossible a generation ago.

And yet the element that makes up roughly sixty percent of our body mass is still treated as something simple and self-evident.

Drink more water.

Stay hydrated.

This advice is repeated so often it has turned into white noise.

But what if hydration is not simply a matter of volume?

What if the most important thing about water is not how much you drink, but how well it matches the biological context in which it is received?

Because water is not generic.

It is shaped by geology. Its character is defined by time. It absorbs minerals as it passes through layers of rock over decades — sometimes centuries. Some waters emerge dense and structured. Others emerge light, soft, and nearly transparent in their mineral profile. Each is a geological expression — as specific and unrepeatable as the rock formation that authored it.

Every spring has its own story. Every source carries its own chemical signature.

And your body can tell the difference.

> Research by the European Food Safety Authority (EFSA) has shown that the bioavailability of calcium and magnesium from mineral water is comparable to — and in some cases exceeds — their absorption from dairy products. Minerals in water arrive in dissolved ionic form; the body does not need to extract them from a solid food matrix. It simply receives a ready signal.

For decades, the culture of health has centered on nutrition, supplements, and efficiency protocols. We learned to read ingredient lists, to argue about fats and carbohydrates, to design personalized vitamin regimens.

But when the subject turns to water, most people think in only two categories: tap or bottled, still or sparkling.

This is roughly equivalent to describing all food with the word "calories."

Reality is more nuanced.

The minerals dissolved in water affect its taste, its absorption rate, its cellular balance, the function of the nervous system, and muscle performance. Some waters feel soft, almost weightless. Others feel structured, grounded, even anchoring.

You may not have the exact words to describe it.

But your physiology understands the difference.

Your body does not simply receive water.

> *"It receives a signal."*

This book begins with a simple idea that changes the way you see:

> *"Water possesses intelligence — not mystical, but geological."*

Not as a form of consciousness — but as the result of a natural "program" created by pressure, temperature, rock, and time.

Think of minerals as an alphabet.

Their proportions as a language.

And your body speaks it fluently.

Long before wearable devices, biohacking, and personalized health laboratories, the human body evolved in constant dialogue with natural mineral water. The body's internal chemistry adapted to it. Its cells learned from it.

> The human body cannot synthesize minerals on its own. All twenty-one mineral elements essential for life — from calcium, which governs muscle contraction, to zinc, which participates in over three hundred enzymatic reactions — must come from external sources. Throughout millions of years of Homo sapiens evolution, water remained one of the earliest and most consistent channels for this delivery. Our receptors, our kidneys, our intestinal lining — all are calibrated for interaction with water carrying geological mineral signatures. This is not a preference. It is evolutionary memory.

When the water changes, so does the conversation.

We live in an age of purification and standardization. Water is filtered, softened, equalized, optimized for large-scale systems. From the standpoint of safety, this is an immense achievement of engineering. Access to safe water is one of the pillars of modern civilization, and it has saved millions of lives.

But in the pursuit of uniformity, something subtle was lost:

Character.

Nature never made water the same twice.

Glacial water carries one mineral signature — typically soft, light, and minimal, formed through slow meltwater with brief geological contact. It is the whisper of the spectrum — not inferior to the denser waters of limestone or volcanic rock, but categorically different. The body reads each distinctly.

Deep artesian springs carry a different story entirely. Years and decades under pressure in limestone or dolomite saturate water with calcium and magnesium, giving it structure and weight.

Volcanic rock creates yet a third profile — rich in silicon and bicarbonates.

These are not marketing narratives. This is geochemistry — the science of water's interaction with the lithosphere over millennia.

And the scientific community is only beginning to understand how significant these differences are for human health.

The future of health is not built on extremes.

It is built on precision.

Personalized medicine. Adaptive nutrition. Contextual optimization of the body. It stands to reason that hydration — the process that underlies all the rest — is entering a new phase as well.

> *"We are approaching an era in which the central question is no longer 'Am I drinking enough?' but 'Does the water I drink match the way I want to function?'"*

This book is neither a critique of water supply systems nor a search for the "perfect" water. It is, rather, an invitation to see water more clearly — to recognize it not as a passive liquid, but as an active participant in human physiology.

Once you understand this, something shifts.

You stop drinking automatically.

You begin to choose deliberately.

And in that choice lies a quiet form of power — the ability to align one of the body's most fundamental resources with the life you want to live.

Because when you change the water, you don't just change what you drink.

You change the conversation happening inside you every day.

Welcome to The Silent Element.

CHAPTER 2

The Simplification Paradox

In 1945, the United States Food and Nutrition Board published a recommendation that outlived all its authors: "A suitable allowance of water for adults is 2.5 liters per day." After a comma, it added a clarification: "Most of this quantity is contained in prepared foods."

The second half of the sentence was forgotten. The first half became dogma.

And so was born one of the most enduring myths in the history of public health: the eight-glasses rule.

The greatest mistake in the modern approach to water is not that we drink too little.

It is that we don't think about what we drink.

The advice to "drink more water" sounds sensible and safe. It has survived decades of health recommendations and become part of the global wellness culture. Doctors repeat it, trainers repeat it, apps repeat it, even smartwatches repeat it.

Over time, this advice acquired the status of an axiom — a statement requiring no further elaboration.

But every axiom has a side effect:

It closes the conversation before it has a chance to begin.

When quantity becomes the only metric, quality inevitably fades into shadow. This is how oversimplification is born. And alongside it — the illusion of understanding.

The human body, however, is far more subtle than the arithmetic of glasses suggests.

We do not pour water into ourselves the way gasoline goes into a tank. The body is not a reservoir, not a system of pipes. It is a dynamic biological environment in which millions of balancing processes occur every second.

> Water constitutes 73% of the mass of the brain and heart, 83% of the lungs, 79% of muscle and kidney tissue, and 64% of skin. But the crucial point is not the percentages themselves — it is that the distribution of fluid between cells and intercellular space is regulated with surgical precision. Even a 2% disruption in water balance — a threshold at which you may not yet feel thirsty — can reduce cognitive function by 20%, according to research from the University of London published in 2018.

Water does not simply exist within the body's processes.

It participates in them.

It influences electrolyte distribution, the function of cell membranes, the transmission of nerve impulses, thermoregulation, and the blood's ability to transport nutrients.

> *"Water participates in every biochemical decision the body makes. It is not the background of biology — it is its co-author."*

And so the question of hydration is inevitably more complex than it appears.

For a long time, universal guidelines were necessary. They helped establish basic habits and prevented the obvious risks of dehydration.

But universality always has a limit.

What works as mass advice rarely serves as a precise instrument.

In recent years, medicine has been gradually moving away from one-size-fits-all solutions. Nutrition is becoming personalized, therapy accounts for genetic differences, even physical training is adapted to individual physiology.

Against this backdrop, the idea that a single hydration formula suits everyone without exception is beginning to look less and less convincing.

Context determines everything.

Climate. Activity level. Diet. Metabolism. Age. Even the nature of your stress. The body is in constant flux — and its needs shift with it.

So why has water remained outside this conversation for so long?

Probably because it seemed too simple for serious attention.

But the simplicity of water is largely an optical illusion.

We see a transparent liquid and automatically assign it neutrality. It seems as though water is always the same — interchangeable, devoid of individuality.

Nature, as is often the case, is more intricate.

> *"Water is a substance that never exists in absolute isolation. Upon encountering the environment, it inevitably enters into interaction. It dissolves minerals. It acquires chemical characteristics. It changes its structure."*

It becomes the result of the conditions through which it has passed.

And those conditions cannot be reduced to zero.

Even the most meticulously purified water is not a return to "primordial emptiness" — it is simply one of many possible forms of treatment. Reverse osmosis, distillation, ion exchange — each method produces water with a distinct profile, a distinct character, and a distinct mode of interaction with the body.

> *"The neutrality of water is a technological assumption, not a natural reality."*

If you take this thought seriously, the familiar question "Am I drinking enough?" gradually transforms into another:

What kind of water is my body actually receiving each day?

How easily does it integrate into the body's internal processes?

Does it demand additional effort to regulate?

Does it support the body's natural balance?

The best work a biological system does is the work you never notice.

Good hydration rarely feels like an event. It manifests as an absence of friction — a state in which the body is not distracted by unnecessary adaptation.

Water simply becomes part of the background against which the body functions freely and precisely.

> *"Hydration is not replenishment. It is the alignment of water with the body. With its current state. With its demands. With its rhythm of life."*

We intuitively understand this logic with food: light meals are appropriate in some situations, heavier ones in others. The body is always signaling what suits it.

Water works by the same principles.

There is no single "correct" water for all occasions. But we can speak of how softly and naturally a given water enters into dialogue with a particular person's physiology.

This thought does not complicate hydration.

It makes it more precise.

And precision is almost always a sign of deeper understanding.

In 2005, the New England Journal of Medicine published a study that upended assumptions about hydration. Of 488 Boston Marathon runners, 13% suffered from hyponatremia — a dangerous drop in blood sodium. The cause was not dehydration. They had drunk too much water. One runner collapsed at the finish line: her sodium had fallen to 121 mmol/L, against a normal range of 135–145. She had been drinking every fifteen minutes, following the advice to "drink more." Her body was not grateful for the maximum. It was asking for precision.

Few health recommendations have proven as durable as the advice to drink more water.

It sounds reasonable. Safe. Even caring.

There is nothing radical about it — and so it is easy to trust.

Over time, this recommendation has become almost a cultural reflex. We hear it from childhood, see it in medical articles, fitness programs, and mobile apps. It has become so habitual that it no longer invites questions.

But here is the paradox.

The advice we repeat most often is the advice we are most likely to mistake for a finished truth — even when it merely marks the threshold of a deeper conversation.

Drinking water is, of course, necessary. That much is beyond dispute.

The question is something else entirely:

Is the idea of "more" sufficient for understanding what the body actually needs?

> The famous "eight glasses a day" rule has no scientific basis. In 2002, Dr. Heinz Valtin of Dartmouth Medical College published a systematic review in the American Journal of Physiology and found no research supporting this recommendation. Its origin likely traces to a 1945 report by the U.S. National Nutrition Board, which mentioned 2.5 liters of daily fluid with a crucial caveat: a significant portion comes from food. The caveat was forgotten. The number survived. And became dogma.

Volume is the simplest parameter to measure. Liters, glasses, milliliters create a feeling of control and clarity. Numbers are reassuring: they turn bodily care into a task that can be "completed."

But the human body rarely submits to arithmetic alone.

It lives in a rhythm of change.

In the heat, the body loses more fluid. During exercise, metabolism accelerates. Stress alters hormonal balance and electrolyte distribution. Even sleep quality can change the next day's water requirements.

> A study published in the journal Sleep (2018) found that people who slept fewer than six hours were 59% more likely to be dehydrated than those who slept eight hours. The reason: disrupted production of vasopressin, an antidiuretic hormone whose secretion peaks during the later stages of sleep. Short sleep literally causes the body to lose more water.

Context is always in motion — and our understanding of hydration must move with it.

And yet universal formulas continue to dominate. Probably because simplicity always looks persuasive.

There is an old intellectual trap: we tend to confuse simple answers with accurate ones. But simplicity is only a first approximation of understanding. It is useful as long as it helps avoid extremes. And it becomes a limitation when it replaces thought.

At a certain point, the advice to "drink more water" accomplished something enormous — it brought attention back to a basic need. But today, as our understanding of physiology deepens, that single guidepost is no longer enough.

A mature relationship with health almost always begins where universal instructions end.

There is one more reason why universal advice is sounding less convincing.

The human body was not designed for extremes. Its resilience is built on balance — on the ability to maintain a stable internal environment despite a changing external world.

Water plays a central role in this process. It helps maintain blood volume, supports thermoregulation, facilitates the transport of substances, and influences the electrical activity of cells.

But precisely because of this, what matters is not just the presence of water, but how harmoniously it enters the system.

Sometimes "more" does not mean "better."

> Hyponatremia — a condition in which blood sodium concentration falls below 135 mmol/L — can result from excessive water consumption. Marathon runners and

> athletes who drink significantly more fluid than they lose are especially susceptible. A study published in the New England Journal of Medicine (2005) found that 13% of Boston Marathon participants showed signs of hyponatremia after the race. The body does not reward maximums. It rewards accuracy.

Sometimes "more precise" means wiser.

> *"Context is the new health. Not universality. Not maximization. But correspondence. The correspondence of water to the body. To its state. To its surroundings."*

Once this becomes obvious, the conversation about water stops being a conversation about quantity.

It becomes a conversation about the quality of interaction.

And that is an entirely different level of attention — quieter, but deeper.

CHAPTER 3

The Mineral Code

In the Italian town of San Pellegrino Terme, nestled in a Lombard valley at the foot of the Alps, there is a spring known since the thirteenth century. Leonardo da Vinci mentioned it in his notebooks. By the nineteenth century, people traveled from all over Europe — not for the landscape, but for the water. Passing through dolomite rock at depths exceeding seven hundred meters, it absorbed calcium, magnesium, and bicarbonates in proportions the body accepted as its own. People did not know the numbers. But their bodies knew the difference.

If you imagine water in its most abstract form, it is easy to picture a transparent, nearly weightless liquid — something universal, without individuality.

But in nature, water is rarely empty.

Almost always, it carries the imprint of the environment through which it passed. Moving through layers of earth, in contact with rock, subject to pressure and time, it gradually dissolves minerals — as though gathering its own vocabulary along the way.

These minerals do not make water complicated.

They make it expressive.

In a sense, it is the minerals that transform water from the chemical formula H_2O into part of a natural context — a substance with a biography.

> *"Minerals are the way water tells the story of where it came from."*

The human body, like any living system, exists in constant exchange with its environment. We are used to thinking of minerals primarily as components of food — magnesium, calcium, sodium, potassium.

But long before nutrition became a subject of scientific analysis, water was already one of the natural carriers of these elements.

Not the primary one. Not the only one.

But a remarkably constant one.

And it is precisely that constancy that makes it significant. What interacts with the body daily almost always exerts a greater influence than rare and intense interventions.

In this sense, water is the most patient medium for the mineral dialogue between the body and nature.

WHAT KEY MINERALS IN WATER DO — A SCIENTIFIC OVERVIEW:

Magnesium (Mg^{2+}) participates in more than six hundred enzymatic reactions in the body. It is essential for ATP synthesis — the primary energy "currency" of cells. Magnesium deficiency is linked to heightened anxiety, muscle cramps, sleep disturbance, and headaches. According to the World Health Organization, a significant portion of the population in developed countries receives magnesium below the recommended intake (310–420 mg per day for adults). Mineral water containing more than 50 mg/L of magnesium can cover 20–25% of daily needs.

Calcium (Ca^{2+}) is not merely a "bone builder." It is critical for nerve impulse transmission, muscle contraction, and blood clotting. A study published in the Journal of the American College of Nutrition (2000) found that calcium from mineral-rich water is absorbed no less effectively than calcium from milk — and for people with lactose intolerance, mineral water becomes one of the few accessible sources.

Bicarbonates (HCO_3^-) form the body's buffering system. They help maintain the acid-base balance of blood, whose pH fluctuates within a narrow range of 7.35–7.45. A shift of even 0.1 pH units can lead to serious disruptions in enzymatic activity. Bicarbonate waters have been used in balneology for centuries to support digestion — and modern science confirms their ability to neutralize excess stomach acidity and improve intestinal motility.

Silicon (Si) is among the least discussed yet important trace elements. It participates in collagen and elastin synthesis, and research (including studies published in BMC Musculoskeletal Disorders) links its presence in drinking water to improvements in skin, hair, and connective tissue health. Some researchers view silicon as a potential protective factor against aluminum accumulation in the body.

Sodium (Na^+) and Potassium (K^+) are the primary electrolytes governing cellular water balance. Their ratio regulates intracellular and extracellular fluid volume, affecting blood pressure, cardiac function, and kidney performance.

But the mineral alphabet tells only half the story. Water with a naturally light mineral profile — the soft, barely perceptible waters that emerge from young aquifers and glacial formations — possesses its own distinct physiological character. Such water is absorbed rapidly, places minimal demand on the kidneys' regulatory systems, and integrates into the body's fluid balance with almost no osmotic friction. For the body that is already well-nourished, this metabolic lightness is not an absence of value. It is a different kind of value — the value of a solvent that performs its most essential function without asking the body to work for it.

It helps to think of minerals not as a set of numbers on a lab report, but as an alphabet — a limited number of symbols from which an infinite variety of "texts" can be composed.

Change the proportion and you change the character.

Some waters feel soft and almost weightless.

Others feel more structured, sometimes even expressive on the palate.

This difference does not call for judgment in terms of "better" or "worse." A more accurate way to speak of it is as different modes of interaction with the body.

Just as in language what matters is not the individual letters but their combination, in water what counts is not merely the presence of minerals but their ratio.

You might call this mineral architecture.

> *"And like all architecture, it can be light or massive, transparent or dense. Each serves a different context. The best is the kind most precisely suited to its purpose."*

For a long time, the conversation about water leaned toward quantitative extremes. Sometimes it seemed as though the higher the mineralization, the more obvious its value — as if concentration automatically equaled benefit.

But deep understanding almost always refuses linear scales.

In biological systems, "more" does not always mean "better." More often, it simply means "different." And "different" is precisely the word the body uses — responding to each mineral profile with a distinct pattern of absorption, processing, and integration.

The body continuously maintains an internal balance — a delicate equilibrium of electrolytes and fluids known as homeostasis. This state is not static; it demands constant work from regulatory mechanisms.

And here an important, though not always obvious, thought emerges:

Sometimes the best support is the kind that does not ask the body for additional effort.

Water can slip gently into this balance, barely drawing the attention of regulatory systems. Or it can require more active adaptation — for instance, when excess sodium forces the kidneys to work harder to excrete the surplus.

Neither scenario is inherently wrong. But the difference between them helps us see hydration more accurately — as a process of interaction, not simple saturation.

> *"Good hydration is not something you feel. It reveals itself as an absence — absence of tension, absence of compensation, absence of the body fighting for balance."*

When the body is not occupied with adaptation, it has more resources for what truly matters: recovery, mental clarity, resilience.

> *"The ideal water does not try to announce itself. It simply lets the body be itself."*

Understanding mineral architecture does not complicate everyday choice. On the contrary — it makes it calmer. We no longer need to chase extremes or submit to abstract scales.

It is enough to ask a more precise question:

How naturally does this water become part of my internal environment?

From that moment, hydration ceases to be a mechanical habit.

It becomes a form of attentive care — a process in which body and water meet without unnecessary resistance.

And perhaps it is in this very invisibility that one of the most underappreciated forms of self-care lies hidden.

In Tokyo's Ginza district, there is a bar that serves only water. No cocktails. No tea. Water — roughly forty varieties from around the world. Each comes with a card: mineralization, pH, mouthfeel, food pairing suggestions. Many dismissed it as eccentricity. But the bar is still operating. Because visitors discover something unexpected: water has a taste. And not just one. Soft Icelandic water at 62 mg/L mineralization feels different from a structured German water at 2,500 mg/L. The first is like a light morning breeze. The second is like a handshake.

We rarely think about the taste of water.

Unlike food, where subtlety and nuance have long been subjects of thoughtful conversation, water in popular perception remains almost tasteless — and therefore, it seems, characterless.

But the expression "tasteless" is deceptive.

More often, it indicates not the absence of taste, but its delicacy.

> *"Noticing a whisper requires more attention than hearing a shout."*

This is precisely why water is one of the most underappreciated sensory experiences in our lives.

We are accustomed to thinking of taste as subjective — a matter of preference. But from a physiological standpoint, taste is first and foremost a recognition system. It is one of the oldest survival mechanisms, allowing the body to rapidly assess the properties of what enters it.

Long before laboratories existed, the body already knew how to discern. Not by analyzing — by feeling.

> The human tongue contains roughly ten thousand taste receptors, organized into taste buds capable of distinguishing five primary flavors — sweet, sour, salty, bitter, and umami. But there is more. In 2019, researchers at the University of California, San Francisco (UCSF) discovered that bitter taste receptors — the TAS2R family — respond to the mineral composition of water and can influence the overall perception of its "softness" or "hardness." In effect, the tongue functions as a miniature chemical laboratory, analyzing the mineral profile of water in fractions of a second.

When water touches the tongue, a process begins that we almost never think about. Receptors instantly respond to dissolved substances, sending signals to the brain. These signals are not always consciously registered, but they participate in shaping our overall sensation — lightness or density, softness or expressiveness.

You could say that taste is the first form of dialogue between water and the body. A dialogue that happens faster than thought.

> *"Taste is not decoration. It is the way the body acquires preliminary knowledge."*

In nature, there is almost no such thing as perfectly neutral water. Even the most delicate springs possess a barely perceptible mineral signature — so faint it is easy to mistake for absence.

But it is often this very subtlety that the body perceives as naturalness.

Strong signals command attention. Gentle ones create a feeling of belonging, as though the body recognizes something familiar.

We may not put it into words, but the body almost always gravitates toward what does not require excess interpretation.

In this sense, taste is not merely a sensory characteristic.

It is an indicator of how easily water integrates into the internal environment.

With the development of gastronomic culture, humanity has learned to value nuance. We discuss the finish of wine, the acidity of coffee, the texture of olive oil. We have come to understand that subtlety does not mean weakness — often, it speaks of complexity and balance.

> In the world of professional water tasting — yes, this field exists — experts differentiate water along several key parameters: mineralization (from ultra-light to concentrated), "texture" (smooth, silky, structured), aftertaste (clean, dry, slightly sweet), and even "weight" on the tongue. The Fine Water Society has developed a classification system in which waters are categorized much like wines — by terroir, mineral profile, and organoleptic characteristics.

Something similar is happening with water, though less visibly.

We are beginning to distinguish not just "like it" or "don't like it," but a calmer question:

How natural does this sensation feel?

Sometimes the taste that is almost impossible to describe is perceived as the most harmonious — not distracting, not overwhelming, allowing water to perform its primary function without unnecessary presence.

> *"The finest water does not try to impress. It tries to be appropriate."*

There is a curious paradox: the cleaner and calmer the sensory experience, the less inclined we are to analyze it — and the more likely it is to be physiologically right for us.

The body values predictability. It does not need to spend resources decoding signals that feel natural.

This is partly why the feeling of lightness after water often proves more informative than any verbal commentary.

The body rarely applauds good decisions.

It simply starts working more freely.

> *"Sometimes the most reliable sign of good water is the absence of any desire to think about it."*

Our era is gradually learning to trust subtle sensations. After decades of fascination with extreme flavors and maximal characteristics, attention is beginning to shift toward balance.

Not louder — more precise.

Not more — more fitting.

Perhaps that is why this philosophy of invisible precision is finding expression in how we experience water. Water accompanies us through all states — movement and stillness, work and recovery. It should not distract. Its role is to sustain the continuity of internal processes.

And here, taste becomes less a sensation and more a form of calibration.

> *"Perhaps the maturity of taste begins the moment we stop seeking impressions and start valuing harmony."*

CHAPTER 4

The Biography of Water

In 2013, geologists from the University of Toronto discovered a pocket of water in the Timmins mine, 2.4 kilometers beneath the surface of the Canadian Shield, that was 2.6 billion years old — more than half the age of the Earth. It was not dead: it contained dissolved hydrogen and traces of reactions suggesting the possibility of microbial life — in total darkness, beneath kilometers of rock, across entire epochs. The water we drink is, of course, younger. But even "young" springs can contain water that went underground thousands of years ago. Between the moment a raindrop vanishes into a crack in the rock and the moment it emerges from a spring, entire ages of human history may pass.

We are used to thinking of water as something instantaneous.

An open tap. A filled glass. A sip taken.

Between the source and this moment, there seems to be almost no distance — in time or in space. Water appears to have no past, as though it has always existed in exactly the form in which we receive it.

But in reality, every water has a history.

Sometimes a history so long that a human life seems, against its backdrop, a brief episode.

Water knows how to wait.

> The age of groundwater can range from several years to tens of thousands. Radiocarbon (^{14}C) and tritium dating show that some artesian sources contain water that entered the earth 10,000–25,000 years ago — during the last Ice Age, when mammoths still roamed Europe. This water filtered through rock formations over the entire span — layer by layer, mineral by mineral — more slowly than mountains grow.

Before reaching the surface, many underground waters spend years, decades, and sometimes far longer in slow transit through the earth. This path is rarely straight. It passes through layers of rock, natural filters, zones of pressure and temperature change.

Each such contact leaves a trace — not in the metaphorical sense, but in the physical.

Dissolved minerals, barely perceptible shifts in composition, structural stability — all of it forms gradually.

Not instantly.

> *"Time is the co-author of water. Every mineral it carries is a record of its passage through the geology of the planet."*

In a world accustomed to valuing speed, it is hard to truly grasp the significance of slow processes. We live in an age of acceleration: information moves instantly, decisions are made quickly, results are expected immediately.

Nature works differently.

Its most enduring forms are almost always born slowly.

Mountains do not appear in a season. Forests do not grow in a year. Coral reefs form over centuries.

And water, taking the long underground path, is part of this same logic.

> *"Slowness in nature is not weakness. It is the method by which stability is achieved."*

The longer water interacts with the geological environment, the more likely its characteristics become stable.

And stability is one of the most undervalued forms of quality.

> Stability of mineral composition is one of the key criteria for certifying natural mineral waters in Europe. Under EU Directive 2009/54/EC, water can be designated as "natural mineral water" only if its composition remains stable over at least two years of regular monitoring. This is not bureaucratic formality — it is an acknowledgment that compositional stability is an objective indicator of source protection and geological maturity.

We have long learned to value time as a factor of quality in other domains. Aged cheeses, wines, fermented foods — in all these examples, duration of process is understood as part of nature's craftsmanship.

With water, this conversation is only beginning.

Perhaps because its transparency prevents us from seeing the labor hidden behind its simplicity.

But the absence of visible effort does not mean the absence of effort. Often, what is cleanest is what took the longest to form.

> *"Water does not keep memories. But it carries the consequences of its journey — a geological biography written*

in its composition. No mysticism. No symbolism. Simply physical reality."

When such water appears in our glass, we rarely consider the scale of time it contains. And yet, in a certain sense, we are touching a substance whose history vastly exceeds our own.

This perspective can change our relationship to the very act of drinking. It ceases to be mechanical. A dimension of connection appears — almost philosophical, yet entirely material.

We are not simply drinking water.

We are entering into contact with its history.

"Sometimes the highest quality is the capacity to remain unchanged."

Observing natural processes, it is hard not to notice a recurring pattern: everything formed without haste tends toward balance. Not toward maximum values. Toward sustainable ones.

Balance is rarely spectacular. It does not try to impress. But it is what allows systems to endure.

The human body is structured the same way. Its health depends not on peak states but on the ability to maintain internal stability.

Water whose own history is bound to a stable environment fits naturally into this logic. Without pressure. Without excess presence.

Understanding the biography of water does not require us to become experts in geology. Rather, it invites a quieter gaze at familiar things.

"In a world obsessed with speed, stability becomes a new form of luxury."

CHAPTER 5

Water and the Brain

In 2012, neuroscientist Matthew Kempton of King's College London asked adolescents to take cognitive tests twice: once after a normal day, and once after drinking a glass of water ninety minutes before the test. Reaction times improved measurably. MRI scans revealed that the brains of the "non-drinking" group worked harder — activating more regions to solve the same problems. The extra energy was not being spent on thinking. It was being spent compensating for a lack of water. Water did not make them smarter. It removed the obstacle.

We are accustomed to linking energy with obvious sources — sleep, food, movement. We understand that insufficient rest makes us slower, and that unbalanced meals affect concentration.

Water almost always remains in the shadows of this conversation.

Its role seems too fundamental to warrant separate discussion. It is perceived as a condition for functioning, rather than a factor capable of noticeably influencing our state.

And yet it is hard to find a process in the human body that does not depend on sustained water balance.

The body's energy begins far earlier than we typically realize.

At the cellular level, water is not merely a medium. It participates in nutrient transport, helps maintain blood volume, influences thermoregulation, and provides the conditions for biochemical reactions.

But perhaps something else matters even more.

Water helps the body maintain internal continuity — the state in which billions of processes occur without disruption or unnecessary strain.

We rarely notice this continuity, because it feels like the norm.

And yet it is one of the most precise definitions of energy.

> *"Energy is not only force. Often, it is the absence of internal resistance."*

Sometimes even a barely perceptible level of dehydration is enough for the body to start working less freely. Concentration drops, fatigue arrives earlier, attention becomes harder to sustain.

These changes can be so subtle that we do not connect them to water. We blame workload, weather, circumstances.

The body rarely sends loud signals. More often, it simply stops being as efficient as it could be.

This is exactly why sustained water balance is not so much a question of survival as it is a question of everyday quality.

But cognition is not the only thing that shifts.

There is a quieter effect of mild dehydration that is, in some ways, more consequential — because it touches not what we think, but how we feel.

Research consistently shows that even modest fluid loss is accompanied by changes in mood: increased tension, a lowered threshold for irritation, a vague sense of unease that has no obvious cause. In the University of Connecticut studies, participants who lost just over one percent of body fluid reported not only fatigue and difficulty concentrating, but heightened anxiety and a diminished sense of well-being — effects that were particularly pronounced in women.

We rarely attribute a bad mood to a glass of water we did not drink. The connection seems too small, too mechanical, to account for something as personal as an emotional state.

And yet the brain does not distinguish between "important" and "minor" disruptions to its operating environment. It responds to all of them. When the internal medium shifts, even slightly, the systems responsible for emotional regulation are among the first to register the change — not with a clear warning, but with a subtle alteration in tone. The day feels heavier. Patience thins. Small frustrations land harder than they should.

Most of the time, we never trace it back.

> *"The mood we inhabit is shaped, in part, by conditions we never think to examine."*

But the conversation about energy is not solely about fluid volume. As with other aspects of physiology, what matters here is not just sufficiency, but how naturally water integrates into the body's work.

The less effort required to maintain internal equilibrium, the more resources remain for other tasks — recovery, adaptation, clear thinking.

You could say that good hydration does not add energy directly. It frees it.

> *"Sometimes the best way to feel more energy is to remove what is quietly taking it away."*

The nervous system is particularly sensitive to water balance. The brain — an organ that constitutes roughly 2% of body mass but consumes 20% of all energy — requires a stable aqueous environment for accurate signal transmission.

> In a study at the University of Connecticut (2012), researchers found that even 1.5% fluid loss — a level at which thirst is barely perceptible — was sufficient to increase fatigue, impair working memory, and heighten anxiety during sustained cognitive tasks. The effect was observed in participants at rest, not during exercise. The authors concluded that mild dehydration increases "cognitive load" — the brain diverts part of its resources to compensating for the altered internal environment instead of directing them at the task itself.

Even small fluctuations can affect cognitive clarity — that ability to think calmly and sequentially, without unnecessary strain.

We often seek clarity through stimulation: caffeine, quick breaks, attempts to "pull ourselves together." But at the foundation of sustained focus there are almost always more fundamental conditions.

One of them is water.

Not as a source of instant lift, but as a factor that sustains stability.

> *"Clarity rarely comes from acceleration. More often, it comes from equilibrium."*

There is one more dimension of this relationship that deserves attention, because it connects two processes the modern world treats as separate: hydration and sleep.

A study published in the journal Sleep (2018) found that adults who slept fewer than six hours were significantly more likely to be inadequately hydrated than those who slept seven or eight. The mechanism is not behavioral — it is hormonal. Vasopressin, the hormone that signals the kidneys to conserve water, reaches its peak secretion during the later stages of sleep. When sleep is cut short, that peak is truncated. The body loses more fluid overnight. The person wakes already in mild deficit.

And then the cycle compounds. Mild dehydration, as the research above demonstrates, degrades cognitive function and emotional regulation during the day. Degraded cognitive function and elevated stress, in turn, are among the most reliable disruptors of sleep quality the following night.

The brain does not experience these as separate problems. It experiences them as a single environment — one in which the conditions for clear, sustained function are either present or absent. Water participates in both halves of that equation: in the quality of waking attention and in the physiological architecture of rest.

> *"The body does not compartmentalize the way we do. What we call 'sleep' and 'hydration' and 'focus' are, for the body, a single conversation about the conditions of its own continuity."*

The human body generally strives not for peak states, but for sustainable ones. Extreme bursts of energy are almost always short-lived. True vitality manifests differently — as the ability to remain functional over extended periods.

In this sense, energy is closer to endurance than to a flash.

Water supports precisely this form of resilience. It participates in cooling during movement, helps maintain circulating blood volume, and contributes to more precise muscle function.

But perhaps its greatest contribution remains invisible — it helps the body not get distracted by the fight for balance.

When internal systems are not overloaded with adaptation, the body begins to function more freely.

And freedom is one of the most accurate sensations of energy.

> *"True energy rarely comes from outside. It appears where nothing stands in the body's way."*

Perhaps one of the most valuable habits is to stop treating water as a secondary detail of the day. Not to wait for serious thirst, not to treat drinking as a mechanical action between tasks.

But to see it as a form of fundamental calibration. Not complicated. But essential.

Over time, this changes the very feeling of life. More stability appears, fewer sharp dips, and attention becomes clearer.

Not because water has magical properties.

But because the body begins to work under the conditions for which it was originally designed.

CHAPTER 6

The Value We Forgot

In April 2018, the residents of Cape Town — South Africa's second-largest city — stood in lines with jerry cans. The city was approaching "Day Zero" — the moment the taps would be shut off. Four million people prepared to live on twenty-five liters a day. Day Zero did not arrive — drastic conservation measures and late rainfall saved the city. But during those months, residents underwent an experience impossible to forget. One woman, an engineer, later told reporters: "I used to turn on the tap without a thought. Now I hear every drop. Not from anxiety. From respect."

Some things are so familiar we stop perceiving them as valuable.

Air. Light. Silence.

And, of course, water.

We rarely think about it as long as it is available — as long as it is enough to open a tap or reach for a bottle. Water seems part of the natural order of things, something almost guaranteed.

> *"Familiarity has a strange property: it renders even the most essential things invisible."*

This is precisely why the true value of water often escapes our attention.

For most of human history, access to clean water was not a matter of convenience, but of survival. Settlements formed near springs, rivers determined trade routes, and droughts altered the fates of entire regions.

In a sense, civilization has always followed water.

> The oldest cities in the world — Jericho, Ur, Mohenjo-daro — were founded near major water sources. The Roman Empire built more than 400 aqueducts spanning over 450 kilometers — one of the greatest engineering achievements of antiquity — to deliver water to cities far from rivers. Control over water resources was a political instrument: whoever controlled the water controlled the territory.

But the modern world has gradually created a sense of distance between humans and this resource. Infrastructure learned to deliver water almost invisibly — and in that invisibility lies one of the great triumphs of the technological age.

We stopped thinking of water as a fragile condition for life.

And began to treat it as a given.

> *"The truest luxury is that which we never have to worry about."*

Yet value is rarely defined by habit alone. More often, it becomes evident at moments of change — when what seemed boundless begins to demand attention.

> According to UNESCO (2023 report), 2.2 billion people worldwide lack access to safe drinking water. Global

> consumption has grown sixfold over the past century and continues to increase by roughly 1% annually. The World Economic Forum has repeatedly listed the water resource crisis among the top five global risks. Moreover, the "water footprint" of daily life is far larger than most people realize. Producing one kilogram of beef requires about 15,000 liters of water. One cotton T-shirt — 2,700 liters. A cup of coffee — 140 liters.

Today, the conversation about water is gradually moving beyond domestic convenience. Urban growth, climate change, rising consumption — all of it compels us to look at water more carefully.

Not with alarm. Not with catastrophic scenarios.

With a new level of understanding.

This shift is happening not only at the level of governments and environmental strategies. It is becoming part of a broader cultural movement — a return of attention to fundamentals.

We are relearning to value sleep. Quiet. Mental balance. Food quality.

Water takes its natural place in this line — not as a trend, but as a foundation.

Water is one of the quietest architects of our internal space. It helps maintain temperature, transport substances, and create conditions for countless biochemical processes.

And it does all of this without any desire to be noticed.

Perhaps this is exactly why we so easily underestimate its role.

> *"The most important supports are rarely at the center of attention."*

Perhaps one mark of wisdom is the ability to see value before it comes under threat. This is not a matter of anxiety, but of clarity.

Caring for water is not merely an ecological gesture. It is a form of long-range thinking, a recognition of the connection between personal well-being and the condition of the environment of which we remain a part.

> *"Civilizations may differ in many ways. But they are united by their dependence on a single source of life."*

CHAPTER 7

The Invisible Cost

In 1993, a team of Finnish epidemiologists at the National Public Health Institute in Helsinki published a study that revealed something surprising about the relationship between water and health. The researchers had spent years analyzing health records from communities across Finland — a country with significant regional variation in water hardness, the geological measure of calcium and magnesium concentration. Their finding: populations drinking soft water — low in calcium and magnesium — showed measurably higher rates of cardiovascular mortality than those drinking harder, mineral-rich water.

The finding was consistent across age groups and could not be fully explained by diet, smoking, exercise, or socioeconomic status.

The study was published in a respected journal. It joined a growing body of similar evidence from Taiwan, the United Kingdom, Sweden, and the United States — research spanning decades. And then, by the standards of public health, something remarkable happened.

Nothing.

No policy review. No public conversation about how the mineral composition of drinking water might interact with health outcomes — and under what conditions.

The evidence did not disappear. It accumulated quietly in academic databases, cited by specialists, unknown to the public.

There is a temptation, when encountering this research, to draw a simple conclusion: mineral-rich water is good, mineral-poor water is bad.

That conclusion would be wrong. Not because the evidence is fabricated — it is not. But because the studies tell a more nuanced story than a linear hierarchy of mineral density would suggest.

The populations studied in these epidemiological investigations share a common characteristic that is rarely foregrounded in popular summaries: they are populations in which *water is a primary source of daily mineral intake.* Communities with limited dietary diversity — where fresh vegetables, dairy, nuts, and other mineral-rich foods are scarce or unaffordable — depend on water for a significant fraction of their calcium and magnesium. For these populations, the mineral content of water is not a refinement. It is nutritionally essential.

For populations with adequate dietary mineral intake — the majority of educated, health-conscious consumers in the developed world — the relationship between water mineral content and health outcomes is far less clear. When calcium and magnesium are supplied sufficiently through food and supplementation, the marginal contribution from water becomes a smaller variable in a larger equation.

This does not make the epidemiological evidence irrelevant. It makes it *contextual* — which is exactly the point this book has been making since its first page.

The invisible cost is not that modern water lacks minerals. The invisible cost is that we stopped paying attention to what our water contains — and therefore lost the ability to recognize when its composition matters, for whom it matters, and in what circumstances.

To understand why this loss of awareness has consequences, it helps to understand how the body processes sustained misalignment between input and need.

In the late 1990s, neuroscientist Bruce McEwen at Rockefeller University developed a concept he called allostatic load — the cumulative biological cost of the body's efforts to maintain internal stability in the face of ongoing stressors. The concept is subtle but profound. Every minor disruption — a night of poor sleep, a skipped meal, prolonged sitting, chronic mild dehydration, a mismatch between what the body receives and what it requires — is, individually, insignificant. The body adapts. The regulatory systems compensate. Life continues without apparent consequence.

But the body keeps a running account.

Each adaptation costs energy. Each compensation requires metabolic work. And over time — months, years, decades — the accumulated toll of these micro-adjustments begins to degrade the very systems responsible for maintaining them.

> McEwen's research demonstrated that allostatic load is not a metaphor. It is measurable — through cortisol levels, inflammatory markers (C-reactive protein, interleukin-6), blood pressure variability, glycated hemoglobin, and other biomarkers. High allostatic load is independently associated with accelerated cellular aging, increased risk of cardiovascular disease, metabolic syndrome, and cognitive decline. The critical insight is that the damage comes not from any single stressor but from the relentlessness of many small ones, sustained over time. The body does not break under acute pressure as readily as it wears under chronic, unrelieved strain.

Misaligned hydration — water that does not match the body's current needs — fits this pattern. Not because any single glass of the "wrong" water causes harm. But because the body's regulatory systems work more efficiently when the inputs they receive are contextually appropriate.

A person whose diet is mineral-poor, who does not supplement, and whose only source of calcium and magnesium is their drinking water — that person is measurably affected by the mineral content of what they drink. The epidemiological evidence is clear on this point.

A person whose diet is nutritionally complete, who supplements strategically, and whose primary hydration need is clean, efficient absorption throughout the day — that person may be better served by water that integrates without metabolic friction. Water that is light, osmotically gentle, easy for the kidneys to process, and free of the regulatory overhead that higher mineral loads can impose.

Neither choice is universally superior. One is appropriate in one context. The other is appropriate in another. The intelligence — the real intelligence — lies in knowing which context you are in.

> *"The cost of misalignment is rarely dramatic. It accumulates in the space between what the body needs and what it receives — quietly, persistently, without announcement."*

The cardiovascular evidence deserves examination in its full scope, precisely because its implications are more nuanced than they first appear.

Since the 1960s, epidemiological studies from dozens of countries have investigated the relationship between water hardness — a proxy for mineral content, primarily calcium and magnesium — and cardiovascular mortality. The pattern that emerges is real and consistent.

> A comprehensive review published in the *European Journal of Cardiovascular Prevention and Rehabilitation* analyzed the accumulated evidence from studies conducted across Europe, North America, and East Asia. The aggregate finding: populations consuming soft water showed cardiovascular mortality rates approximately 10–25% higher than those consuming harder, mineral-rich water. The finding persisted after adjustment for standard confounders including diet, physical activity, and smoking.

Individual studies sharpen the picture. In Taiwan, a study published in *Environmental Research* in 2006 drew on health records from over three million residents and found a significant inverse relationship between magnesium levels in drinking water and death from coronary heart disease. In Sweden, researchers at the Sahlgrenska University Hospital demonstrated that women living in areas with the highest calcium concentrations in water had a 25% lower risk of dying from heart disease compared to those in areas with the lowest. In the United Kingdom, the British Regional Heart Study found that cardiovascular mortality was 10–15% higher in soft-water regions of England and Wales — a finding that controlled for social class, smoking, and alcohol consumption.

These are not marginal studies. They are large-scale, population-based investigations published in peer-reviewed journals over more than fifty years.

But — and this distinction is critical — they are studies of *populations*, not of individuals. They measure aggregate outcomes across communities with varying diets, economic conditions, and supplementation habits. They demonstrate that water mineral content matters *as part of a total nutritional context.* They do not demonstrate that any individual drinking low-mineral water, within a nutritionally adequate diet, faces measurably elevated cardiovascular risk.

The mechanisms are well understood. Magnesium is essential for vascular smooth muscle relaxation, cardiac rhythm stability, and modulation of inflammatory pathways. Calcium participates in blood coagulation, endothelial function, and vascular tone regulation. When water delivers these minerals in dissolved ionic form — the form in which the body absorbs them most readily — it provides a persistent, low-dose supplementation that operates below conscious awareness. For people whose diets are mineral-insufficient, this contribution is physiologically meaningful. For people whose mineral intake is adequate from other sources, it is one input among many.

The word for what these studies reveal is not crisis.

The word is *context.*

The skeletal evidence follows a parallel logic — and a parallel qualification.

Calcium is the most abundant mineral in the human body. Approximately 1.0–1.2 kilograms of it reside in the bones and teeth of an adult — serving not only as structural material but as a reservoir the body draws upon when circulating calcium levels fall below the narrow range required for nerve transmission, muscle contraction, and blood clotting.

Bone is not inert. It is living tissue, continuously remodeling throughout life — osteoclasts disassembling old bone, osteoblasts laying down new. The balance between these processes determines skeletal density. In youth, formation outpaces resorption. In later life, the balance reverses. The rate at which it reverses depends, in part, on the sustained adequacy of calcium intake over the preceding decades.

> The WHO report on the health significance of water mineral content — published initially in 2004, with ex-

> panded commentary in 2005 — noted that for populations relying on water treated to remove mineral content, there is elevated risk of mineral insufficiency when other dietary sources are limited. The report was authored by a panel of researchers including Frantisek Kozisek of the Czech National Institute of Public Health, and it drew on epidemiological data from multiple countries. The authors found that for people with limited dietary diversity — which includes not only low-income populations but also the elderly, the young, and anyone on a restricted diet — drinking water can supply up to 20% of daily calcium and 25% of daily magnesium. For these populations, the mineral profile of water is not a marginal variable. It is a meaningful dietary input.

The body does not signal a marginal calcium shortfall with a specific symptom. It adjusts. It draws more calcium from the skeletal reservoir. It downregulates processes that can tolerate marginal reductions. The individual feels nothing. The bones record everything.

And here a critical point emerges: the skeletal consequences of sustained marginal calcium insufficiency do not manifest in youth, when the body's compensatory capacity is greatest. They manifest in middle age and beyond — as reduced bone density, increased fracture risk, accelerated osteoporosis. By the time the damage is measurable, the decades of inadequate input are already inscribed in the architecture of the skeleton.

This is the particular cruelty of subclinical depletion. The cost is paid long after the cause has ceased to be identifiable. And the solution, for a given individual, depends entirely on their total calcium intake from all sources — food, supplements, and water. For those whose diet provides adequate calcium, the water's mineral contribution is supplementary. For those whose diet does not, it can be significant.

The neurological dimension introduces a similar pattern — and, again, a similar dependence on individual context.

Magnesium participates in over six hundred enzymatic reactions in the human body. Among its roles: the synthesis of ATP — the body's primary energy molecule. The regulation of neurotransmitters, including serotonin, GABA, and glutamate. The modulation of the hypothalamic-pituitary-adrenal (HPA) axis — the body's central stress-response system. The maintenance of synaptic plasticity — the brain's ability to form, strengthen, and reorganize neural connections.

Chronic subclinical magnesium deficiency does not produce an identifiable disease. It produces a cluster of symptoms that are individually unremarkable and collectively debilitating: heightened anxiety, disrupted sleep, difficulty sustaining concentration, muscular tension, headaches, increased sensitivity to stress, a persistent background sense of fatigue that sleep does not fully resolve.

These are the symptoms of modern life. They are attributed to overwork, to screen exposure, to insufficient exercise, to the pace of contemporary existence.

They are almost never attributed to total magnesium intake — and water's role within it is rarely examined.

> According to data compiled by the National Institutes of Health (NIH) and corroborated by surveys across Europe and Australasia, an estimated 50–80% of the population in developed countries does not meet the recommended daily intake of magnesium (310–420 mg for adults, varying by age and sex). The causes are multiple: industrial agriculture has depleted magnesium from topsoil; processed and refined foods diminish intake further. Water that carries 30–100 mg/L of magnesium can supply 60–200 mg daily, covering 15–50% of

> requirements — a meaningful contribution for anyone not meeting intake through diet alone. For people who are meeting their magnesium needs through food and supplementation, the water's magnesium content is less critical — and other qualities of the water, including its absorption efficiency and metabolic lightness, may be more relevant to daily wellbeing.

The question is not whether magnesium deficiency affects neurological function. That question has been answered extensively across decades of neuroscience research. The question is whether an individual's total magnesium intake — from all sources, including water — is adequate for their physiology. For some people, mineral-rich water closes a genuine gap. For others, the gap does not exist, and the water's primary value lies in other qualities.

This is the central insight that the existing research points toward, once the linear hierarchy of "more minerals is always better" is set aside: water's biological role is contextual. Its value depends on the person who drinks it.

> *"The gap is not between what we know and what we ignore. It is between the institutions that regulate safety and the institutions that study nutrition. The water falls through the space between them — and the individual falls through the space between the studies."*

There is a temporal dimension to this story that deepens its significance.

The human body's relationship with water was established over approximately 300,000 years of Homo sapiens' existence — and over millions of years of hominid evolution before that — in constant interaction with natural water sources: springs, rivers, lakes, aquifers filtered through geological strata. These waters varied enormously in their mineral profiles. The soft, light glacial meltwaters of northern latitudes. The dense, calcium-rich springs of Mediterranean limestone. The silica-laden streams flowing through volcanic basalt. The human body did not evolve to expect a single mineral concentration. It evolved to interpret a range — to process geological diversity, to adapt its absorption and excretion to whatever the local water carried.

The body's kidneys, intestinal lining, electrolyte regulatory systems, and cardiovascular function were all shaped within this context. Not designed for a specific mineral concentration — designed for *variety*, and for the ability to calibrate its response to whatever input arrived.

What changed in the twentieth century was not, primarily, the mineral content of water. What changed was the *uniformity*. For the first time in the species' history, billions of people began drinking water of a single, standardized character — determined not by geology but by infrastructure. Whether that water was mineral-low (as in many treated municipal supplies) or mineral-enriched (as in some remineralization programs), the defining shift was from geological diversity to industrial consistency.

The body did not lose access to minerals. It lost access to the diversity of inputs within which its regulatory systems operate most naturally — the varied geological conversation that shaped its calibration.

> *"Evolution did not calibrate the body for mineral-rich water or mineral-light water. It calibrated the body for diverse water — water whose character varied with geography, season, and source. The mismatch is not between the body and a mineral level. It is between the body and the uniformity."*

Emerging research in epigenetics introduces an additional dimension — one that must be approached with appropriate scientific caution, but that cannot responsibly be omitted from a serious discussion of this topic.

Epigenetics studies heritable changes in gene expression that do not involve alterations to the DNA sequence itself. Environmental factors — nutrition, stress, chemical exposure, and other aspects of what epidemiologists call the exposome — can modify how genes are activated or silenced. Crucially, some of these modifications appear to be transmissible across generations. The child may carry, in its gene expression patterns, the biological record of the parent's environment.

The question, not yet conclusively answered but actively investigated in multiple research programs, is whether chronic inadequacy of total mineral intake — from all sources, including diet, supplementation, and water — can produce epigenetic effects. Can sustained low calcium intake alter the expression of genes governing bone metabolism? Can chronic magnesium inadequacy modify the epigenetic regulation of the HPA stress axis? Can these modifications, once established, persist into the next generation?

> Research published in the *American Journal of Clinical Nutrition* and in *Epigenetics* has demonstrated that maternal mineral status during pregnancy — particularly calcium, magnesium, and zinc — influences fetal bone development, kidney maturation, and metabolic programming. These effects are well established for dietary minerals. The Dutch Hunger Winter study and similar investigations of prenatal nutritional stress have shown that the metabolic consequences of inadequate mineral intake during pregnancy can persist for decades and may be heritable. Whether water's mineral contribution

> plays a measurable role in this equation depends, as with the cardiovascular and skeletal evidence, on the total nutritional context of the individual. For a person whose diet and supplementation provide adequate minerals, water's contribution to the epigenetic equation is a smaller variable. For a person whose total mineral intake is marginal, every source — including water — becomes more significant. The exclusion of this variable from the research is itself notable. It tells us something about the limits of what the current research architecture considers worth measuring.

This is the frontier of the inquiry. It is not settled science, and this book does not claim it is. But it is a question that demands investigation — because its implications, if confirmed, extend the timescale of the conversation considerably.

If total mineral adequacy is not only a concurrent factor but a cumulative one — one whose biological consequences deepen across decades and may transmit across generations — then the question of whether an individual's mineral intake is contextually appropriate acquires a longer horizon than a single lifetime.

This does not argue that any particular water is dangerous. It argues that total mineral awareness — understanding where your minerals come from and whether the sum is adequate — is a question worth taking seriously. The answer, as always, depends on context.

Why has this conversation remained so quiet?

The question is not rhetorical. The answer is structural, and understanding it is essential to understanding why the relationship between water composition and health has received so little public attention.

The studies cited in this chapter are not obscure. They are published in major peer-reviewed journals — the *New England Journal of Medicine, Environmental Research*, the *European Journal of Cardiovascular Prevention*, the WHO's own technical publications. The epidemiological evidence linking water mineral content to health outcomes has been documented for over fifty years, across dozens of countries, with remarkable consistency.

The silence has three structural causes.

First, the findings are context-dependent. The health effects of water mineral content are most pronounced in populations with limited dietary mineral intake. For populations with nutritionally adequate diets, the individual effect is smaller and harder to isolate. This makes the research genuinely difficult to translate into universal recommendations — and easy to set aside in a clinical culture that prefers actionable, individual-level advice.

But the existence of a contextual variable does not make it unimportant. It makes it worthy of understanding.

Second, the achievement that created the modern water supply is real and sacred — and rightly so. Modern water treatment saves lives. Millions of them, every year, without interruption. Cholera, typhoid, dysentery — diseases that devastated human settlements for millennia — were conquered by the very processes that also altered the mineral composition of water. Any conversation about water composition that does not begin by honoring this achievement deserves to be dismissed. The point is not that water treatment is wrong. The point is that its nutritional implications were never examined as a variable worth understanding.

Third, no institution is responsible for understanding the full picture. Water treatment is regulated for the *presence* of contaminants — pathogens, heavy metals, chemical pollutants, disinfection byproducts. Nutritional science studies food. The mineral composition of drinking water inhabits the corridor between these two disciplines — visible to researchers in both, studied systematically by neither, owned by no one.

This is not a conspiracy. It is a bureaucratic architecture in which an important question fell through the gap between adjacent jurisdictions. And it is a gap that only contextual awareness — at the individual level — can begin to close.

> *"We solved the acute problem — waterborne disease — with an engineering triumph that has saved millions of lives. The contextual question — what does my water contain, and does it match what my body needs? — was never assigned to anyone. It remains, by default, yours to answer."*

The invisible cost is not a catastrophe.

It is something quieter and, in its way, more consequential — the progressive narrowing of awareness about a substance that constitutes sixty percent of the human body. Not a mineral crisis, but a perceptual one. A civilization that stopped thinking about water — its character, its composition, its geological identity — and reduced the entire relationship to a volume metric.

It is the cost of a simplification that was never named, never debated, never presented to the public as a choice.

The evidence that water composition affects health exists. It has existed for decades. It is published, peer-reviewed, and available to anyone who searches for it. What the evidence shows is not that one type of water is universally superior. It shows that *context determines outcome* — that water's biological effect depends on who is drinking it, what else they are consuming, what their body needs in that moment, and whether the composition of their water matches or misaligns with those needs.

What has been missing is not data.

It is the framework for interpreting the data — the understanding that water, like food, like sleep, like exercise, is a contextual input whose value depends on its relationship to the body that receives it.

This chapter is an attempt to provide that framework. It does not claim that any single type of water is the cause of disease. It does not argue for a universal mineral standard. It claims something simpler and harder to dismiss:

That the composition of drinking water is a health variable that deserves to be examined, understood, and chosen with intention.

That the refusal to examine it — the treatment of all water as interchangeable — is itself a form of negligence, not by any institution, but by a culture that has stopped paying attention to the most fundamental substance in human life.

And that naming this negligence — making it visible — is the necessary first step toward the kind of literacy that should have existed all along.

> *"Visibility has always been the precondition for informed choice. You cannot choose well what you have never learned to see."*

CHAPTER 8

The Boundaries of the Body Are Porous

Beginning in the late 1960s, biologist Lynn Margulis argued — and eventually demonstrated — that our very cells are the result of an ancient merger. Mitochondria, the energy stations in every cell, were once independent bacteria. They entered our ancestral cells billions of years ago — and stayed. And in 2016, science journalist Ed Yong reminded us: the human body houses as many bacterial cells as it does its own. We are not a single organism. We are an ecosystem. The boundary of the body is not a wall. It never was.

We are accustomed to perceiving the body as something enclosed — a system separated from the outside world by skin, as though by a natural border between "us" and everything else.

This feeling is so intuitive it rarely faces scrutiny. We speak of inside and outside, of protection, of what should remain out and what may be allowed in.

But from the standpoint of biology, the picture is considerably more complex.

A living organism does not exist in isolation. It is in continuous exchange with its environment — substances, energy, signals. Breathing, eating, the movement of fluids, heat exchange — these are all forms of constant dialogue between the body and the world.

The boundary exists.

But it is permeable.

> Over the course of an average human life, roughly 35,000 liters of water pass through the body. But what is even more striking: the water molecules inside the body are replaced on a cycle ranging from a few days (blood plasma renews approximately every 48 hours) to several weeks (intracellular fluid). Red blood cells live about 120 days. Intestinal epithelial cells, 3–5 days. Skin cells, 2–3 weeks. We are literally renewed. In material terms, we are not the same people we were a few months ago. And water is one of the primary conductors of that renewal.

Every breath alters the chemical composition of the blood. Every meal becomes part of cellular structures. The water we drink soon ceases to be external — it integrates into the complex system of internal balance.

After a while, it becomes impossible to draw a clear line and say where "it" ends and "we" begin.

In a certain sense, the body is a process, not an object.

A process of continuous renewal.

> *"We remain ourselves not because we are unchanging, but because we know how to renew."*

This thought may seem unfamiliar, but it leads to an important understanding:

> *"Life is possible not because of separation, but because of managed permeability. A fully closed system cannot remain alive."*

Water makes this connection especially obvious. It moves freely between the external and internal, participating in temperature regulation, substance transport, and the maintenance of cell volume.

But perhaps its most important role is to remind us that the boundary of the body is not a wall. It is, rather, a fine membrane — a structure that simultaneously protects and allows interaction.

Without this permeability, life would be impossible.

Modern culture often emphasizes the idea of control — the desire to manage the body, optimize it, shield it from unfavorable factors. There is reason in this. But sometimes, alongside control, the illusion of total autonomy appears.

As if the body were capable of existing on its own.

Biology says otherwise.

We depend on the air we breathe. On the food we receive. On the water that becomes part of our internal processes.

This dependence is not weakness.

It is a form of inclusion in the larger system of life.

> *"The autonomy of living organisms is always relative. Connection is their fundamental state."*

The concept of the "exposome," proposed by epidemiologist Christopher Wild in 2005, describes the totality of environmental exposures a person undergoes from conception to the end of life. It includes air, water, food, chemicals, social factors, and stress. Exposome research shows that 70–90% of chronic diseases are determined by environmental factors, not genetics. We are not isolated from the world — we are formed by it.

This continuity changes the optics. Water ceases to be perceived as a merely consumed resource. It becomes part of a broader process — a circulation in which all living things participate.

> *"We interact with the world not from time to time. We interact always."*

This thought extends well beyond physiology. It touches the very way we see ourselves — not as a separate unit, but as part of a continuous system of exchange.

Such a view does not diminish individuality. On the contrary, it makes it realistic.

We remain unique, but not severed. Stable, but not static. Self-sufficient, but not isolated.

> *"We are, to a far greater extent, a process of interaction with the world than a separate form."*

In this understanding there is a peculiar calm. It releases us from the illusion of total independence and at the same time reminds us of a simple responsibility — to be more mindful of what we let into our lives.

Not from anxiety.

From clarity.

CHAPTER 9

The Water Audit

In 2019, a research team at the University of Dublin conducted a study that revealed less about water than about the limits of modern health awareness. The researchers asked two hundred adults — educated, health-conscious people living in one of the world's wealthiest countries — to list the minerals in the water they drank most frequently. Fewer than four percent could name even one.

The same participants could, on average, identify the macronutrient composition of their lunch. They could name three vitamins in their morning smoothie. They tracked their sleep in ninety-minute cycles and knew their resting heart rate to within two beats per minute. Several could describe the specific origin of the coffee beans they used that morning.

They did not know what was in their water.

Not because the information was secret. Because no one had ever suggested it was worth knowing.

This book has made an argument across its preceding chapters: that water carries a geological signature — a compositional identity shaped by time, rock, and pressure — and that the human body responds differently to different signatures. That the modern reduction of water to a generic, interchangeable commodity has obscured a relationship that deserves to be understood. And that the biological relevance of water composition is not universal but contextual — depending on the individual, their diet, their physiology, and their current state.

If that argument is sound, it leads to a practical question that cannot be deferred: what does a person do with this knowledge?

The answer begins not with action but with perception. Before you can align your hydration with your context, you must first see what you are drinking. And for most people, the water they consume every day is, in compositional terms, entirely opaque. They know its temperature. They know whether it is still or sparkling. They may have a preference for one brand over another, based on taste or habit. But the geological character — the actual compositional identity — is invisible to them.

This is not a personal failing. It is a systemic one. Nowhere in formal education, in public health messaging, in medical training, or in consumer culture has anyone built the framework for seeing water as anything other than a transparent, interchangeable liquid.

This chapter builds that framework.

Not by turning you into a hydrologist. Not by prescribing a regimen or ranking waters on a scale of virtue. But by giving you a set of tools — simple, concrete, and applicable this week — that make the invisible legible.

Literacy before strategy. Seeing before choosing. Understanding before optimizing.

> *"You cannot improve what you cannot see. And for most people, the mineral content of their daily water is as invisible as the air they breathe — and far more consequential than they realize."*

WHAT WATER LITERACY MEANS

We have, in recent decades, developed the concept of literacy for many domains beyond reading and writing.

Financial literacy — the ability to understand interest rates, debt structures, and investment risk — is now recognized as a civic necessity. Its absence is treated not as a personal eccentricity but as a systemic vulnerability.

Nutritional literacy — the ability to read food labels, distinguish macronutrients from micronutrients, and evaluate dietary claims — has transformed how millions of people eat. The mandatory nutrition label, introduced in the United States in 1990 and now standard across most of the developed world, was not merely a regulatory act. It was an act of literacy creation. It made the invisible visible.

Digital literacy — the ability to navigate information ecosystems, evaluate sources, and recognize manipulation — is increasingly understood as essential for democratic participation.

Water literacy belongs in this sequence. Not at its periphery, but at its foundation — because water is consumed more frequently, more constantly, and with less deliberation than any other substance.

Water literacy is the ability to read, interpret, and act on the mineral composition of the water you consume.

This is not expertise. A financially literate person is not an economist. A nutritionally literate person is not a dietitian. Water literacy does not require a chemistry degree or a subscription to the *Journal of Hydrology*. It requires the capacity to answer three questions:

Where does my water come from?

What does it contain?

Does that composition support what my body needs?

Most people today can answer none of these with any specificity. Not because they are careless or unintelligent, but because no institution — no school, no public health authority, no medical practice, no consumer advocacy group — has ever systematically framed these questions as important.

They are.

> *"Literacy is not the accumulation of specialized knowledge. It is the acquisition of a question you did not previously know to ask."*

THE WATER AUDIT: A PROTOCOL FOR SEEING

The Water Audit is not a permanent habit to be sustained through willpower. It is a one-time diagnostic — a structured period of observation, lasting approximately one week, designed to reveal the mineral landscape of your daily water intake.

Think of it as a blood test for your water, rather than for your body. A snapshot that, once taken, permanently changes what you are able to see.

It works in five steps.

STEP ONE: IDENTIFY YOUR SOURCES.

For seven days, notice every instance of water entering your body. This includes the water you drink directly — from a tap, a filter pitcher, a countertop filtration system, a bottle purchased at a shop. It also includes the water that enters through food preparation: the water you boil pasta in, brew coffee or tea with, use for soups, rice, grains, and reconstituted foods.

Most people discover, upon examination, that their daily water comes from two or three sources they have never consciously examined. Many discover a source they had not even thought of as a "water choice" — the ice from the freezer dispenser, the water at the office cooler, the tap at the gym.

Write them down. Specificity matters. "Tap water" is not sufficient. The question is: tap water from which municipality? Treated by which process? Sourced from which river, reservoir, or aquifer? These are not arcane details. They are the equivalent of knowing which field your food was grown in — except that, for water, almost no one asks.

STEP TWO: FIND THE COMPOSITION.

This step sounds technical but is, in practice, remarkably simple.

For bottled water, the mineral composition is printed on the label — typically on the back, in small type, often beneath the barcode or near the recycling information. The key values to look for: calcium (Ca), magnesium (Mg), sodium (Na), potassium (K), bicarbonates (HCO_3), and total dissolved solids (TDS). Some labels also report silica (SiO_2), chloride (Cl), sulfates (SO_4), fluoride (F), and pH.

For tap water, every municipal water utility in the developed world publishes an annual water quality report — called a Consumer Confidence Report (CCR) in the United States, a Water Quality Report in the European Union and United Kingdom, and analogous documents in most other countries. These reports disclose mineral content, treatment methods, contaminant levels, and source information. They are publicly available, usually online, at no cost. Most people have never read theirs. Many do not know such a document exists.

For filtered water, the critical question is: what does the filter remove? This question has a precise answer, and it varies dramatically by filter type. Activated carbon filters — the kind found in most pitcher filters and many faucet-mounted systems — primarily remove chlorine, some organic compounds, and certain heavy metals. They generally leave the mineral composition of water largely intact. Reverse osmosis systems, by contrast, remove virtually everything — including calcium, magnesium, and bicarbonates — producing water with a mineral profile close to zero. The distinction between these two types of filtration is significant, and most people who use filters daily do not know which category their system falls into.

> A study published in the *Journal of Water and Health* tested the mineral content of water before and after treatment with common household filtration systems. Activated carbon filters reduced chlorine by 90% while retaining approximately 85–95% of calcium and magnesium content. Reverse osmosis systems removed 90–99% of all dissolved minerals, reducing TDS from an average of 320 mg/L to 12 mg/L. The study's authors noted that consumers were largely unaware of this difference and assumed all filters performed similarly. Each filtration technology produces water with a fundamentally different character — and the right choice depends on whether the individual needs mineral contribution from their water, or values the metabolic lightness and absorption efficiency that lower-mineral water provides.

One distinction deserves particular attention here, because it is often overlooked: naturally low-mineral spring water and industrially stripped reverse-osmosis water may appear compositionally similar on a label — both show low TDS, low calcium, low magnesium. But they are fundamentally different substances. A natural spring with low mineralization has a geological identity — a stable composition certified over years of monitoring, shaped by a specific aquifer, protected by regulatory standards. Reverse osmosis water has no geological origin. Its composition is the product of mechanical removal, not geological formation, and it varies with the source water fed into the system. The numbers may converge. The character does not.

STEP THREE: RECORD THE KEY NUMBERS.

You do not need to become an analytical chemist. Five values are sufficient to characterize any water and to understand what it is delivering to your body:

Calcium (Ca), in mg/L — The body's most abundant mineral. Essential for bone maintenance, nerve impulse transmission, muscle contraction, and blood coagulation. The recommended daily intake is approximately 1,000 mg for most adults.
Magnesium (Mg), in mg/L — Involved in over six hundred enzymatic reactions. Critical for ATP synthesis (the body's energy currency), nervous system regulation, cardiovascular function, and sleep quality. Recommended intake: 310–420 mg daily.
Sodium (Na), in mg/L — The primary extracellular electrolyte. Governs fluid distribution between cells and blood plasma, influences blood pressure, and is lost significantly through sweat.
Bicarbonates (HCO_3), in mg/L — The body's principal buffering system. Maintains blood pH within the narrow range of 7.35–7.45 that is required for enzymatic function. Natural bicarbonate content is a marker of geological origin and is associated with digestive and metabolic support.
Total Dissolved Solids (TDS), in mg/L — An aggregate measure of all dissolved mineral and organic substances. A rough but reliable indicator of overall mineralization. Useful as a first-glance characterization of any water.

> The European Union classifies natural mineral water into categories based on TDS. Below 50 mg/L: "very low mineral content." Between 50 and 500 mg/L: "low mineral content" (sometimes labeled "oligomineral" in Mediterranean countries). Between 500 and 1,500 mg/L:

"medium mineral content." Above 1,500 mg/L: "rich in mineral salts." These categories are not a hierarchy — they are a typology. Each reflects a different geological origin and a different mode of interaction with the body. Understanding the categories is the first step in understanding what you are drinking — and whether it matches what your body needs.

STEP FOUR: CONSIDER CONTEXT.

Once you have your numbers, hold them against two questions — not two value judgments, but two diagnostic considerations.

First: what is my total daily mineral intake from all sources? If your diet is nutritionally complete and you supplement strategically, your body's mineral needs may already be well covered. In this case, your water's primary role is efficient hydration — clean absorption, osmotic compatibility, minimal regulatory overhead. A naturally light water with low TDS performs this role with precision. It integrates without friction. It allows the kidneys to operate in their most efficient range. It supports the body without adding to its processing burden.

Second: is there a gap in my mineral intake that water could help close? For some people — those with restricted diets, limited access to fresh food, or specific physiological needs — water that carries meaningful calcium and magnesium can serve as a quiet, continuous mineral supplement. In these contexts, a higher-mineralization water is not a luxury. It is a practical nutritional contribution.

The composition panel tells you what the water offers. Your context tells you what you need. The match between the two is the definition of intelligent hydration.

A water with Ca 80, Mg 26, TDS 485 offers substantial mineral delivery in dissolved, bioavailable form. A water with Ca 3, Mg 1,

TDS 22 offers metabolic lightness — clean, rapid absorption with minimal renal load. Neither is inherently superior. Each serves a different biological purpose. The intelligent choice depends entirely on the person, the moment, and the context.

STEP FIVE: ASK ONE QUESTION.

Does the water I drink match what my body needs — not in theory, not by default, but right now, given what I know about my diet, my activity, and my physiology?

You do not need to answer with laboratory precision. The question itself is the intervention. What was opaque becomes visible. What was automatic becomes considered. And from that shift in perception — not from anxiety, not from obsession, but from simple clarity — better choices follow naturally.

"The Water Audit does not tell you what to do. It tells you what you have been doing — and lets you decide whether that is enough."

THE HYDRATION ARCHITECTURE: SEEING THE FULL PICTURE

The body does not distinguish between minerals from food, minerals from supplements, and minerals from water. It receives them all through the same absorptive pathways — the intestinal epithelium, the renal tubules, the cellular transport channels calibrated over evolutionary time to process dissolved ions regardless of their vehicle.

But we, as people managing our own health, tend to think in compartments. We track food carefully — often meticulously. We add supplements deliberately, sometimes spending considerable sums on specific formulations and delivery systems. And we ignore water almost entirely, as though it were metabolically inert.

The Mineral Stack is a concept designed to correct this blind spot. It maps your total daily mineral intake from all three sources — food, supplements, and water — and reveals where the gaps are.

The process is straightforward. Estimate your daily dietary mineral intake using any standard nutritional database or tracking tool. Record any supplements you take and their dosages, noting the specific compound (magnesium oxide is absorbed at roughly 4%; magnesium citrate at approximately 25%; magnesium glycinate higher still — the form matters as much as the dose). And add the mineral contribution from your daily water, using the numbers you gathered in the Water Audit.

For most people, the result is clarifying in a specific way.

The Hydration Architecture reveals the relative contribution of each source — food, supplements, and water — and makes visible the structure of your total daily mineral intake and hydration quality. For some people, water is a significant mineral source, contributing 15–25% of daily calcium and magnesium. For others — those with comprehensive diets and thoughtful supplementation — water's mineral contribution is minimal, and its primary value lies elsewhere: in the quality of its absorption, its osmotic compatibility, its role as the body's most constant solvent.

> A study published in the *Journal of the American College of Nutrition* compared calcium absorption from mineral water (containing approximately 170 mg/L calcium) with absorption from milk. The researchers found that calcium bioavailability from mineral water was comparable to — and in several measured parameters exceeded — that from dairy. The explanation: minerals in water arrive already dissolved in ionic form, requiring no digestive processing. This makes mineral-rich water one of the most efficient mineral delivery mechanisms

available — a fact relevant for people whose diets are mineral-insufficient. For people whose diets are nutritionally complete, water's contribution to the mineral economy is a smaller variable, and other qualities — absorption kinetics, renal gentleness, osmotic ease — may matter more in the context of daily hydration.

The Hydration Architecture does not prescribe action. It reveals structure. And what it reveals varies by individual: for some, the water layer is a significant contributor that should be chosen with mineral delivery in mind. For others, the water layer's primary role is the quality of the hydration itself — and the minerals are better sourced from food and supplements already elsewhere in the architecture.

> *"The Hydration Architecture does not tell you which water to drink. It tells you what role water plays in your total mineral economy — and lets you choose accordingly."*

THE HYDRATION ALIGNMENT SPECTRUM: CONTEXT AS COMPASS

There is one final framework that completes the architecture of this chapter, and it addresses the question this book has been building toward since its opening pages: if water quality matters, and if different waters carry different mineral profiles, then how does one choose?

The answer is not a formula. Formulas belong to the paradigm this book is challenging — the paradigm that treats all bodies as identical and all moments as equivalent. The answer is a principle: alignment.

The body's needs are not static. They shift across the hours of the day, across seasons, across levels of activity and recovery. Morning physiology differs from evening physiology. The body after sustained physical exertion differs from the body at rest. The demands of a hot, arid climate differ from those of a temperate, humid one. Even the nature of one's work — cognitive versus physical, prolonged concentration versus intermittent activity — alters the body's relationship to water.

The Hydration Alignment Spectrum is a framework for matching water's mineral profile to the body's current physiological state. Not with clinical precision — that would require continuous monitoring that does not yet exist outside of research settings — but with informed, contextual awareness.

It operates on a simple principle: different states of the body benefit from different mineral inputs, and the most effective hydration is the kind that meets the body where it currently is.

Consider the morning. After six to eight hours of sleep, the body has lost approximately 200–300 milliliters of water through respiration and transdermal evaporation. Vasopressin — the antidiuretic hormone — is declining from its nighttime peak. Cortisol is rising. The kidneys are recalibrating. This is a moment for gentle rehydration: water that the body can integrate without regulatory strain — water whose composition demands nothing the body is not ready to give. High sodium or very high TDS water in this context creates an unnecessary regulatory demand at a moment when the body's systems are still transitioning from rest to activity.

> Research in chronobiology confirms that the body's fluid regulation follows circadian rhythms that are genetically encoded and hormonally mediated. Vasopressin secretion peaks during the later stages of sleep and diminishes after waking. Cortisol, which influences renal blood flow and glomerular filtration rate, follows its own circadian curve with a morning peak (the cortisol awakening response). Aldosterone, which regulates sodium retention and potassium excretion, fluctuates

> throughout the day. The body's receptivity to water — its capacity to integrate a given volume and composition without triggering compensatory mechanisms — is not constant. It varies with the hormonal and metabolic context of the moment. This does not argue for complex protocols. It argues for awareness that context exists.

During sustained cognitive work — the kind that characterizes much of modern professional life — the priority shifts to stability. The brain requires a consistent aqueous environment for accurate neurotransmission. Fluctuations in electrolyte concentration can disrupt the electrical gradients that nerve cells depend upon for signal propagation. Water with moderate mineral content and low sodium supports cognitive stability — providing mineral input without triggering the osmotic adjustments that divert regulatory resources.

During and after intense physical activity, the equation changes again. Sweat is not pure water. It contains sodium (typically 40–60 mmol/L), potassium, chloride, and smaller amounts of calcium and magnesium. Replacing lost volume without replacing lost electrolytes risks dilutional hyponatremia — the dangerous condition documented in the Boston Marathon study that opened this book's second chapter. Water with higher sodium and potassium content is better aligned with the body's post-exertion state.

In the evening, as the body transitions toward sleep, alignment shifts once more. Higher sodium intake in the hours before sleep activates renal filtration, increasing the likelihood of nighttime waking — a disruption that, as earlier chapters have established, impairs vasopressin secretion and compounds the cycle of dehydration. Moderate magnesium, by contrast, supports muscle relaxation and is associated in clinical studies with improved sleep onset latency and sleep quality.

The Hydration Alignment Spectrum is not a rigid prescription. It is a compass — a way of asking a better question than "Am I drinking enough?" The better question: "Is this water, at this moment, aligned with what my body needs?"

> *"Alignment is not optimization. It is the practice of meeting the body where it is — not where a universal formula imagines it should be."*

READING THE LABEL

There is a practical skill woven through everything above, and it is worth addressing directly, because it is the simplest and most immediate application of this chapter's ideas.

Most people have never read a water label. Not the front — the name, the logo, the mountain imagery — but the back. The composition panel. The small-print table that lists, in milligrams per liter, what the water actually contains.

That table is the geological signature made visible. It is the biography of the water, translated into numbers.

Reading it takes fifteen seconds.

You look for five values: Ca, Mg, Na, HCO_3, TDS. With those five numbers, you can understand any water's geological identity — where it came from, what it passed through, what character it carries.

A water reading Ca 80, Mg 26, Na 6.5, HCO_3 357, TDS 485 tells one geological story: extended contact with carbonate rock. Dense mineral architecture. This is water that delivers — calcium, magnesium, buffering capacity. For a person seeking mineral support through water, this profile is highly functional.

A water reading Ca 3, Mg 1, Na 2, HCO_3 12, TDS 22 tells a different geological story: a young aquifer, minimal rock contact, or a formation that produces naturally soft water. This profile is light, clean, and metabolically unobtrusive. The body processes it with minimal regulatory effort. For a person whose mineral intake is adequate through diet and supplementation, and whose primary need is efficient daily hydration, this water performs with quiet precision.

Both waters are expressions of geology. Both are natural. Both serve the body — in different ways, for different purposes.

Once you learn to read this information — and it truly requires no more than a single attentive encounter with a label — you cannot unread it. The composition panel stops being fine print and becomes a geological biography. The water stops being a generic commodity and becomes a specific substance with a specific character and a specific role.

This is not connoisseurship. It is not ranking. It is the most basic form of literacy applied to the substance you consume more frequently and in greater quantity than any other.

> *"You read food labels because someone taught you they mattered. No one has taught you to read water labels. That is the literacy gap. Closing it takes fifteen seconds."*

THE ONE-PERCENT SHIFT

There is a temptation, when encountering new information about health, to overcorrect — to transform insight into obsession, to replace one form of inattention with another form of anxiety, to seek total control over yet another variable in an already over-monitored life.

This book does not advocate for that.

The frameworks presented in this chapter — the Water Audit, the Hydration Architecture, the Hydration Alignment Spectrum — are instruments of visibility, not systems of control. They exist to make the invisible legible, not to add another layer of optimization to a culture already saturated with self-improvement protocols.

The shift that matters is small. A one-percent adjustment in the quality of attention.

You read one label. You check your municipality's water quality report — once. You notice, when choosing between two waters, that their compositions differ — and that the difference is not random but geological. You begin to sense, over weeks and months, that your body responds differently to different waters — not dramatically, but perceptibly. A slight change in how you feel after your morning glass. A barely noticeable improvement in the quality of your sleep when your evening water carries more magnesium and less sodium.

These are not revolutionary events. They are recalibrations — small adjustments in the quality of a relationship that you have maintained, unconsciously, every day of your life.

> Research in behavioral science consistently demonstrates that sustainable health improvements are more strongly correlated with small, perception-based shifts than with large, effortful interventions. A study published in *The Lancet* on dietary behavior change found that interventions focused on increasing awareness and environmental visibility of choices outperformed those requiring significant willpower or behavioral restructuring. The same principle applies here: making water composition visible is a more effective and durable intervention than prescribing a hydration protocol, because visibility operates at the level of perception rather than effort. What you can see, you begin to consider. What you consider, you gradually align.

The philosopher Alfred North Whitehead once observed that civilization advances by extending the number of operations we can perform without thinking about them. There is truth in this. Automation frees attention for higher-order problems. But there is also a cost — because what we stop thinking about, we also stop seeing. And what we stop seeing, we can no longer evaluate, protect, or improve.

This book has been, in its entirety, an invitation to reverse that process for one substance. To take something that had become automatic — so automatic that it was functionally invisible — and return it, briefly, to the field of conscious attention.

Not permanently. Not obsessively. Not with the anxious vigilance of someone who has been told they are doing everything wrong.

Just enough to see what was always there.

> *"You do not need to become an expert. You need to become literate. The difference is one label, one question, one pause."*

The tools exist now. The geological signature can be read. The audit can be completed in a week. The stack can be mapped on a single page. The spectrum can orient your choices without dictating them.

None of this required specialized knowledge. All of it was available before you opened this book.

What changed is not the information.

What changed is the question.

And the question — *what is in the water I drink, and does it match what my body needs?* — once asked, does not easily close.

It stays open.

Quietly.

The way all important questions do.

CHAPTER 10

What Continues in Silence

In the Buddhist tradition, there is a practice called kinhin — walking meditation. The monk takes one step and stops. Feels the foot on the floor. Takes the next step. The purpose is not to arrive somewhere. The purpose is to be present in each step. Imagine if we could relate to a sip of water in the same way. Not to turn every sip into a ritual — but to notice, at least occasionally, the moment itself: liquid touches the lips, moves downward, the body receives it. This moment lasts less than a second. But within it lies one of the most frequent acts of exchange between a person and the world.

Some interactions with the world happen so frequently that we stop being aware of them. Breathing, the change of light through the day, the sensation of air temperature — all of it continuously affects us while remaining almost invisible.

But with nothing external does a person interact as regularly and directly as with water.

Several times a day, we perform the same gesture — take a sip. And in that moment, the line between inside and outside becomes almost notional: what was, just an instant ago, part of the world begins to participate in the body's work.

This process is devoid of drama.
Which is exactly why it is so easy to miss.

> *"The most constant dialogues with life happen without words."*

Human attention is selective. We tend to react to the sharp, the novel, the urgent. Quiet processes are perceived as background — until something disrupts their flow.

And yet the quality of our lives is shaped, in large part, by that very background.

> Neuroscience research over the last decade shows that conscious attention to simple actions — including eating and drinking — activates the insular cortex, the brain region responsible for interoception — the perception of the body's internal state. The more developed one's interoception, the better a person recognizes the body's needs — from thirst and hunger to stress and fatigue levels. Research at University College London (UCL) found that people with high interoceptive sensitivity make more accurate decisions under uncertainty — their bodies literally "suggest" the right choice before the conscious mind arrives there. The ability to notice internal signals is not mysticism. It is a trainable neurobiological skill.

Noticing does not mean analyzing every action. It is, rather, a particular form of presence in which we do not walk past the things that support us constantly.

When attention touches even something as habitual as a sip of water, it ceases to be automatic. A sense of alignment begins to emerge — an almost imperceptible moment of cooperation between the body and its environment.

> *"To notice water is to notice the very moment of exchange with life."*

The body almost always sends signals before overt discomfort appears. A slight dryness, a dip in focus, a barely perceptible tiredness — these are not annoyances. They are forms of feedback.

But hearing them requires a space of attention.

In a culture of acceleration, this sensitivity is gradually dulled. We grow accustomed to responding to the urgent and less and less frequently notice the gradual.

This is why the ability to notice can be called an art — not an innate talent, but a quality that develops alongside inner maturity.

Water has an extraordinary way of returning us to this simple calibration. It accompanies movement and rest, work and recovery, periods of concentration and moments of release. Its presence is so natural that we rarely question the very fact of its constancy.

And yet it is hard to find another example of interaction with the world that repeats so often over the span of a life.

From the first sip in childhood to deep old age, water remains our most continuous contact with the external environment.

> *"We may not notice water, but the body never stops interacting with it."*

Over time, attention begins to change the experience of everyday life itself. We start to distinguish between drinking automatically and drinking in response to a real need, between haste and alignment, between habit and care.

These distinctions are barely visible from the outside. But it is exactly from such distinctions that inner equilibrium gradually forms.

There is a common misconception that attention requires strain. In reality, genuine attention is more like a calm readiness to perceive — without the urge to control everything.

In this state, the body begins to feel less like a mechanism in need of constant correction and more like a living system with its own capacity for self-regulation.

Our task becomes simpler: do not ignore its signals.

> *"Sometimes the deepest form of intelligence is the ability to notice what supports us every day."*

With age, many people come to especially value the basic conditions of life — sleep, gentle rhythms, quality food, clean air, water. Not because they have discovered something new, but because they have learned to see significance in the familiar.

Noticing almost always precedes caring. We rarely protect what we are not aware of.

And the reverse: what enters the field of attention gradually comes to be perceived as value.

> *"What we are able to notice gradually becomes what we know how to care for."*

In 1990, Voyager 1, then six billion kilometers from Earth, turned its camera back at the request of Carl Sagan. In the photograph, Earth was a tiny pale blue dot in a beam of light. Sagan wrote: "That's here. That's home. That's us. Everyone you love, everyone you know — on a mote of dust suspended in a sunbeam." That dot is blue because it is covered in water. But of all the planet's water, 97.5% is saltwater. Of the remaining 2.5%, two-thirds is locked in glaciers. Available freshwater amounts to less than 1%. One percent. An entire civilization — on one percent.

For most of history, water was perceived as the natural backdrop of existence — something so obvious it rarely became the subject of reflection. Settlements arose near springs, rivers guided the movement of civilizations, rains set the rhythms of agriculture.

We have always depended on water.

But we have rarely thought about that dependence.

> *"What accompanies life constantly is the first thing to begin seeming guaranteed."*

The infrastructure of the modern world has made access to water almost invisible. It is enough to turn a tap — and the immensely complex natural and technological processes behind that flow remain outside our awareness.

This is one of civilization's greatest achievements: transforming a life-essential resource into quiet routine.

And yet such comfort has a particular quality — it gradually distances us from understanding the true value of what we receive without effort.

In recent decades, in different parts of the world, a thought has been gaining ground that once sat at the periphery of collective consciousness: the conditions that sustain life require not only technology, but attentive stewardship.

This shift cannot be called sudden. It more closely resembles a slow maturation — the moment when humanity begins to recognize its own embeddedness in the fragile systems that make existence possible.

> According to projections by the UN Water Institute, by 2030 the global freshwater deficit could reach 40% — a gap between the available supply and the needs of population, industry, and agriculture. The distribution of water resources is extremely uneven: Brazil holds roughly 12% of the world's freshwater, while countries in the Middle East and North Africa live under chronic scarcity. Already, some four billion people — nearly two-thirds of the world's population — experience water shortage at least one month per year.

This is not about alarm. It is about clarity.

The value of water reveals itself not only where it is scarce. It is present in the very fact of its existence — in its capacity to sustain health, to support communities, to create conditions for flourishing.

In this sense, water is not merely a resource.

It is one of the fundamental forms of life's continuity.

> *"Water reminds us: what sustains us is always larger than we are."*

The recognition of value rarely arrives in an instant. More often, it forms gradually — alongside the expansion of knowledge, the growth of populations, shifts in climate, the development of technology. All these processes seem to invite us to look deeper and see connections that once went unnoticed.

> The "planetary boundaries" framework, developed by a team of scientists led by Johan Rockström at the Stockholm Resilience Centre (2009), identifies nine key Earth systems whose transgression threatens biospheric stability. Freshwater use is one of them. By these researchers' estimates, humanity is approaching the critical threshold of freshwater extraction from natural cycles, beyond which cascading ecosystem degradation begins — from rivers drying up to the destruction of wetlands that serve as natural filters and climate regulators.

The same water that maintains an individual's physiological balance is part of global cycles — moving through atmosphere, earth, and rivers, returning again and again in a continuous process of renewal.

In a certain sense, every sip connects us to a system that vastly exceeds an individual life.

Perhaps one of the quietest transformations of our time is precisely this — the passage from treating water as an infinite backdrop to understanding it as a value.

Not a scarcity. Not a privilege. But a foundation.

This understanding rarely leads to grand gestures. Rather, it shifts the optics — making our attention slightly keener, our decisions slightly more responsible, and our view of the future slightly more whole.

The maturity of a civilization is revealed, in no small part, by how it treats the things without which it cannot exist. Not only in the ability to extract resources, but in the capacity to preserve the conditions that allow life to continue.

> *"True value reveals itself not through scarcity, but through the recognition of irreplaceability."*

In Japanese, there is a word — komorebi. It means sunlight filtering through the leaves of trees. Not the light itself and not the trees themselves — but precisely their meeting. A moment in which one passes through the other and creates something that belongs to neither.

Perhaps our relationship with water is a komorebi of another kind. We are not water, and water is not us. But in the moment it passes through us, something third emerges: life, which would be impossible without that meeting.

Water is always near us.

It accompanies the morning, when the day is only beginning to unfold, and stays through the evening, when the world gradually slows. We reach for it almost without thinking — between tasks, between words, between thoughts.

This gesture is so simple it easily slips from notice.

And yet there is something in it larger than habit.

Every sip is a reminder of connection. A reminder that life is sustained not only by great events but by processes that unfold continuously and almost invisibly.

Perhaps, over the course of this book, water has gradually revealed itself differently — not only as a physiological necessity, but as part of a broader experience: of interaction, attention, and alignment with what surrounds us.

But understanding rarely requires solemn moments. More often, it arrives quietly — the way dawn comes, without needing to announce itself.

At some point, we simply begin to see a little more clearly.

> *"The world does not slow down. Circumstances do not simplify. What changes is only the quality of our presence in our own lives."*

We begin to notice what we used to walk past. To feel the difference between the automatic and the intentional. To treat with greater care the things that support us every day.

Not as an obligation.

More as a natural form of attention.

> *"Sometimes the most profound changes occur without outward signs — they simply become part of how we live."*

Water does not demand that we think about it. It does not aspire to be a symbol and needs no special significance. Its role is simpler and, at the same time, greater — it continues to flow, participating in the endless movement of life.

Long before us.

And far beyond our individual story.

To realize this is to sense the steadiness of the world — a steadiness that exists independent of our haste.

Perhaps this is why the simple act of drinking sometimes takes on an almost meditative quality. For an instant, the clutter falls away, and all that remains is the immediate experience — coolness, motion, quenching.

Nothing more is needed.

Life is already happening.

"What truly sustains us rarely needs to be loud."

This book does not offer finished answers. It was, rather, an invitation — to look more carefully at what was always nearby.

The rest happens without words.

Each person continues this conversation in their own way — through their own observations, sensations, choices.

Perhaps that is what real understanding looks like: not a period, but a space in which thought can continue to move.

When you next take a sip of water, everything will be as simple as before.

And yet, perhaps, in that gesture a pause will appear — a moment of recognition.

The same glass. The same motion. The same coolness.

But for a fraction of a second, you will feel something you had not noticed before. Not a thought — more a sensation. As though, for an instant, you found yourself at the point where your body and the planet meet. Where your internal environment and the external world turn out to be the same thing. Where fifteen thousand years of underground travel end with a touch against your lips.

This moment lasts less than a second.

But within it is the entire book. The entire conversation. The entire connection.

> *"We think we drink water. But perhaps it is water that drinks us — absorbing us into its current, making us part of a flow older than any language and longer than any memory."*

Books end.
The current does not.
And everything that truly matters continues exactly where silence begins.

A NOTE ON SOURCES

This book draws on peer-reviewed research published across multiple decades in journals including the New England Journal of Medicine, Environmental Research, the European Journal of Cardiovascular Prevention and Rehabilitation, the American Journal of Clinical Nutrition, the Journal of the American College of Nutrition, and the technical publications of the World Health Organization. The epidemiological evidence on water mineral content and health outcomes spans more than fifty years and dozens of countries. Where specific studies are cited, their details are provided in the text. The scientific context sections throughout this book are intended to make the evidence transparent to the reader without the interruption of endnotes. Readers seeking the primary sources will find them readily through the author, journal, and date information provided within each citation.

ABOUT THE AUTHOR

Evgeny Basov is a writer and entrepreneur whose work addresses the relationship between civilization and the resources upon which it silently depends. His central preoccupation is water—not as a commodity or a subject of policy, but as a foundational condition of biological and social life, one that most societies have ceased to think about carefully.

At a moment when considerable energy is directed toward the colonization of other planets, Basov turns attention to a prior question: whether we have understood, or adequately protected, the elemental systems that sustain life on this one. His writing treats water as a lens through which to examine deeper failures of perception—the tendency of societies to overlook what is most essential until it is no longer available.

He is the founder of Prana Spring, a natural water brand, and supports initiatives to expand access to safe drinking water in underserved regions. He is also the founder of PlasticToArt—a project dedicated to collecting plastic and transforming waste into art, approaching pollution as a material capable of acquiring new form and meaning.

His entrepreneurial work and his writing share a single premise: that the way a civilization relates to its most basic resources reveals something essential about its capacity to endure.

Through his writing, Basov engages with questions that precede and outlast any single generation—questions about what we owe to the material foundations of life, and what it means to inhabit a world whose most critical systems remain, for most people, invisible.

www.ingramcontent.com/pod-product-compliance
Lightning Source LLC
LaVergne TN
LVHW090328210926
848480LV00004B/1339
* 9 7 8 9 9 4 8 6 3 4 0 7 2 *